Clouds

By Jane Manners

CELEBRATION PRESS
Pearson Learning Group

Contents

Learning About Clouds 3
Cirrus Clouds 7
Cumulus Clouds 10
Stratus Clouds 13
Different Kinds of Clouds 16

Learning About Clouds

Clouds come in all shapes and sizes.
Clouds are always changing, too.
It's fun to look at clouds.

Clouds are part of the water cycle. In the water cycle, water moves from the Earth to the sky. Then it moves back down again.

Water Cycle

2 The warm air rises and meets colder air. The water vapor condenses, or changes into tiny water drops or ice crystals. The drops and crystals form clouds.

1 The sun warms water and air. The water evaporates, or changes into water vapor.

3 The water drops and ice crystals join together. When they get heavy enough, they fall to Earth as rain, sleet, hail, or snow.

There are different kinds of clouds. Clouds have different shapes. They also form in different parts of the sky.

Cirrus Clouds

Cirrus clouds are feathery and thin. Sometimes these clouds are called "horses' tails." Can you see why?

Cirrus clouds form very high
in the sky where it is very cold.
These clouds are made of ice crystals.

Cirrus clouds may form on a fair day. These clouds can also appear before a storm.

Cumulus Clouds

Cumulus clouds look like puffy cotton balls. They often form low in the sky.

Sometimes cumulus clouds take the shape of common things. What do you see in these clouds?

When cumulus clouds become large and dark, they turn into nimbus storm clouds. This means that rain may be on its way.

Stratus Clouds

Stratus clouds are like damp sheets. They hang low and often cover the whole sky.

When you see stratus clouds, rain or snow may be on the way.

Different kinds of clouds tell us
about different kinds of weather.

Different Kinds of Clouds

cirrus clouds

cumulus clouds

stratus clouds